Campden & Chorleywood Food
Research Association Group

Chipping Campden
Gloucestershire
GL55 6LD UK
Tel: +44 (0) 1386 842000
Fax: +44 (0) 1386 842100
www.campden.co.uk

Guideline No. 38

The Use of Chlorine in Fresh Produce Washing

David Dawson

2002

Information emanating from this
company is given after the
exercise of all reasonable care and
skill in its compilation,
preparation and issue, but is
provided without liability in its
application and use.

Legislation changes frequently.
It is essential to confirm that
legislation cited in this publication
and current at the time of printing,
is still in force before acting
upon it.

The information contained in this
publication must not be
reproduced without permission
from the Publications Manager.

Campden & Chorleywood Food Research Association Group comprises
Campden & Chorleywood Food Research Association
and its subsidiary companies
CCFRA Technology Ltd   CCFRA Group Services Ltd   Campden & Chorleywood Magyarország

© CCFRA 2002
ISBN: 0 905942 54 X

SUMMARY AND SCOPE

Fresh produce is often washed in order to remove particles of dirt or soil from outer surfaces. Removal of microorganisms is also a major objective for those operators washing fruit and vegetables, particularly those using biocides such as chlorine.

There is no specific guideline on the use of chlorine for fruit and vegetable decontamination, in spite of the fact that it is used by the majority of the fresh produce industry in the UK. This document therefore sets out to do two things. Firstly, it provides useful background information on the use of chlorine for washing fresh produce and some of the wider issues associated with its use. In doing this it also reports some data from CCFRA, obtained using a pilot scale wash tank. Secondly, it advocates, on the basis of current knowledge, some broad principles of best practice for chlorine-based washing of fruits and vegetables.

Use of chlorine requires understanding of the basic principles. The principles should ensure that chlorine is used in the most economical, safest and most effective way. There are no guarantees on specific degrees of inactivation of bacteria with particular levels of chlorine, but following the principles of this guideline maximises the chances of the chlorine being effective.

Current industry practices on chlorine use are outlined, summarising a CCFRA review of industry practice published in 1999. This Guideline describes the principles of chlorine disinfection, and explains the significance of pH and measurement of free chlorine levels. It also deals with the principles of chlorine measurement in the factory. Case studies carried out at CCFRA on different product types with a purpose built, highly controlled 300-litre pilot-scale tank are used to demonstrate the effects that organic loading has on chlorine levels. Decontamination is also illustrated using these pilot scale conditions. The CCFRA equipment offers an example of a controlled washing system and some of the practicalities of running such systems are discussed, including some problems that may arise. Specific engineering aspects of washing tank design and the design of chlorine control systems are, however, beyond the scope of this document.

Legal issues surrounding the use of chlorine are discussed, together with perceptions about its desirability and safety. Alternatives to chlorine are reviewed briefly, but not in depth.

The key generic principles and "take-home messages" emphasised in this guide are:

- Both controlled and uncontrolled chlorine-based disinfection systems can function effectively if managed properly.

- It is essential to characterise a washing system for a particular operation - for example, in terms of volume and type of produce.

- Periodic manual checks that chlorine targets are being achieved are recommended, even for systems where this is done automatically.

- Adequate training of operatives is essential to the running and full effectiveness of any washing operation.

- It is not essential to maintain high levels of chlorine (>100ppm) throughout the washing process in order to achieve disinfection, but it is important to know how much chlorine is required to achieve the required efficacy in the system.

- Loss of free chlorine can compromise disinfection.

In essence, this guide provides the background facts necessary to understand the basis of chlorine-based disinfection and provides prompts and suggestions on the factors that need to be considered in establishing and running an effective system.

CONTENTS

		Page no.
1.	INTRODUCTION	1
2.	DESCRIPTION OF CCFRA WASH TANK AND CHLORINE DOSING AND CONTROL SYSTEM	4
3.	CURRENT PRACTICE	6
4.	PRINCIPLES OF CHLORINE DISINFECTION	8
5.	ORGANIC LOADING - PRINCIPLES AND CASE STUDY	11
6.	CONTROL OF CHLORINE AT LOW LEVELS: PRINCIPLES AND CASE STUDY	14
7.	WHAT TO MONITOR: FREE CHLORINE, TOTAL CHLORINE OR REDOX POTENTIAL?	18
8.	ALTERNATIVES TO CHLORINE	20
9.	CONCLUSIONS	23
10.	ACKNOWLEDGEMENTS	24
11.	REFERENCES	25
Appendix 1	Chlorine Measurement	26
Appendix 2	Method for spiking and recovering microorganisms from vegetables	28

1. INTRODUCTION

The increase in sales of raw, ready-to-eat produce in recent years has given rise to questions about the removal of microorganisms from such produce and also the efficacy of washing systems.

It is known that fruit and vegetables may become contaminated with microorganisms which may cause spoilage of produce or may be human pathogens.

The contamination rate of fresh produce with particular pathogens is likely to reflect the country of origin, the conditions in which the crop is grown and harvested and possibly the method used for study. In a survey of imported lettuces by the Public Health Laboratory Service (PHLS), none of the following pathogens were found from the following in 151 samples analysed: *Salmonella*; *Shigella*; *Campylobacter; Listeria monocytogenes* and *Escherichia coli* O157 (Little et al. 1999). More recently, however, *Salmonella* Newport was linked to commercially produced lettuce in the UK both epidemiologically and through analysis of samples (Anon 2001).

In the USA, an FDA study of domestically grown produce carried out in 2000-2001 showed that 12 of 1028 samples were positive for *Salmonella* or *Shigella* (Lash 2002). The produce types were chosen because they are normally eaten raw.

Washing processes are designed to remove dirt and soil, insects and foreign material. In addition, most processors are looking for some microbiological reduction in their washing process. This may be to reduce spoilage, but is also to remove any pathogens which may be present. HACCP plans may assume some removal of pathogens at the washing stage, but the prevention of contamination is the best method of ensuring food safety. Fruits and vegetables may become contaminated with pathogenic microorganisms whilst growing in fields or orchards, or during harvesting, post-harvest handling, processing and distribution. A polluted cultivation environment, or poor hygiene in processing increases the risk of contamination with food-borne pathogens.

Given that minimising contamination is one way to prevent foodborne illness, decontamination still has a role to play. It is not clear how effective the decontamination used in the industry is, because there are no data on removal of naturally occurring pathogens and no information on how significant these pathogens may be epidemiologically. In spite of this uncertainty, if a decontamination system is being employed, it is useful to understand how to maximise its effectiveness.

A range of decontamination methods are in use in the fresh produce industry: these include washing in potable water or in water containing chlorine, chlorine dioxide or ozone. Alternatively organic acids, surfactants, or a combination of treatments may be used. There is no list of permitted processing aids in the UK and EU, although an advisory inventory adopted by the Codex Alimentarius Commission at its 18th session in 1989 includes a number of microorganism control and washing agents.

Any processing aid which has biocidal activity falls under EC Directive 98/8/EC (the "Biocides Directive"). Only biocidal products containing an active substance which has been approved under the Directive will be authorised for use: chlorine is authorised under this directive. However, existing products within the scope of the Directive will only be able to to stay on the market until the active substances they contain have been reviewed under the new rules. The Directive gives 10 years from May 2000 for all existing active substances to be reviewed.

In 2001, the Food Standards Agency instructed the Meat Hygiene Service and the corresponding body in Northern Ireland to enforce fully the prohibition on the use of hyperchlorinated water in the production of poultry meat. Before this time, the use of such waters had been permitted on public health grounds despite the fact that UK and EU legislation only allows the use of potable water in poultry meat production. It could be implied that treatment of this water by a food business with large amounts of chlorine may render it non-potable, but the precise legal position in the UK is not clear. It may be argued that any additional treatment of water after it enters a factory or food premises renders it non-potable whether this treatment be with chlorine or other oxidative disinfectants, or indeed organic acids.

Chlorine has had some negative coverage recently and parts of the food industry would like to reduce or obviate the need for its use. It is worth noting, however, that the water industry has gone through some of these issues in the past with concerns over disinfection byproducts such as trihalomethanes, but in most countries it is still key to water treatment regimes and the WHO still regards it as critical to the safety of water worldwide (Anon 1996).

In spite of the above facts and although chlorine is banned for treatment of organic produce, it and potable water alone were the most commonly used methods of fresh produce washing in 1999 as determined by CCFRA in a survey of the industry (Seymour, 1999). At this time, 76% of respondents used chlorine, whilst 20% used potable water alone. It is appropriate that this guideline concentrates upon the use of chlorine in comparison with potable water alone as a decontamination method.

Advice on the use of chlorine in fruit and vegetable washing is limited. The conclusion of the WHO publication "Surface decontamination of fruits and vegetables eaten raw: a review" (Beuchat, 1998) states that concentrations of 200ppm chlorine generally reduce populations by 10-100 fold and that vigorous washing with water alone can be as effective as this. It also states that the population of microorganisms decreases as the chlorine concentration is increased to about 300ppm, above which effectiveness is not proportional to increased concentration. CCFRA has identified a wide range of chlorine concentrations in use within the industry, with levels up to 600ppm being reported.

For effective disinfection of water for drinking the WHO recommends a level of free chlorine of 0.5 mg/litre (ppm) or greater for at least 30 minutes at below pH 8.0 (Anon, 1996). It has however been recognised for many years that microorganisms attached to surfaces are much more resistant to disinfection than planktonic microorganisms (Sommer *et al*, 1999). Logically this would apply to plant surfaces as much as inert surfaces; therefore the same level of kill cannot be achieved in vegetable decontamination.

In order to demonstrate some of these factors, CCFRA built a 300 litre pilot scale washer with an integral safety cabinet. This equipment was used to carry out decontamination work under a range of different conditions and also to look at the effects of different vegetables upon chlorine demand.

Following on from this work, the aim is not to provide a prescriptive document or to go into the engineering detail of washing systems, but to identify key facts.

2. DESCRIPTION OF CCFRA WASH TANK AND CHLORINE DOSING AND CONTROL SYSTEM

The CCFRA pilot scale washing system (Fig. 1) comprises a stainless steel washing tank linked to a dosing and monitoring system. Water is drawn off by a pump near to the upper water level. The outlet is protected by a mesh filter to ensure that fine particles do not get into the monitoring equipment or pumps. The circulation loop is completed as water is returned into the bottom of the tank. In this loop there are two ports for addition of solutions of citric acid or sodium hypochlorite. These solutions mix with water as they reach the bottom of the tank. The outgoing part of the circulation loop is sampled prior to dosing of citric acid and sodium hypochlorite for total chlorine, free chlorine, redox potential and pH. The sample lines pass through a dedicated probe for each parameter being measured. A dilution system comprising a header tank was employed for measurement of higher levels of total chlorine beyond those for which the probe was designed. A similar system was

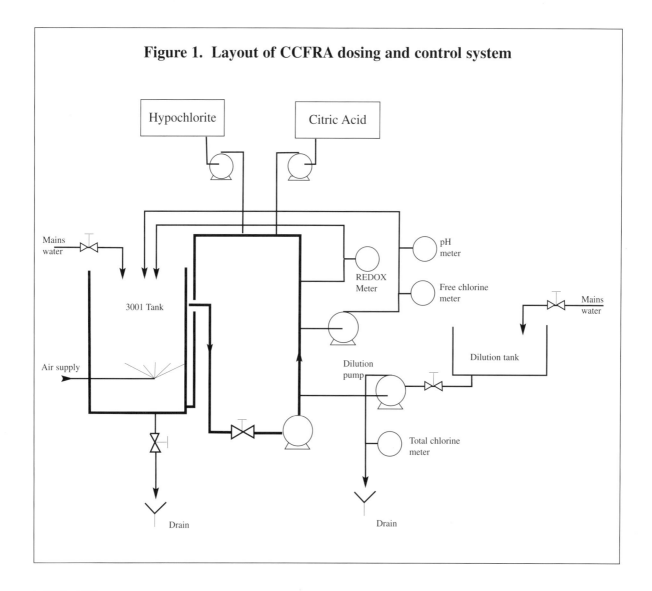

Figure 1. Layout of CCFRA dosing and control system

employed for the free chlorine meter, but this was removed when work was carried out using lower chlorine levels (5 and 10ppm).

Air agitation was possible using filtered compressed air fed into the bottom of the tank. Pasteurisation of the tank was also possible via a steam fed heat exchanger at the bottom of the tank (not shown in Fig.1).

The pilot fresh produce washing tank at CCFRA can be used as a batch washer without control. The chlorine level is brought up to a particular point by use of the dosing pump, which is then switched off; the pH is similarly adjusted by use of citric acid dosing, before the produce is introduced. Alternatively it can be used as a controlled system whereby a chlorine level is maintained at a set point due to sensing of chlorine level and automatic dosing. Chlorine level can thus be automatically readjusted as organic matter causes it to drop. Measurement may be on line or manual. Various combinations of these methods were used during the course of the experimental work, but the following points are relevant to most systems and are worth noting:-

- It is easy with such a system to exceed the set point: this is due to the relatively small size of the system giving little buffering capacity. This is particularly likely to be an issue if strong solutions are used. By the time they have properly mixed in and the levels in the tank have been monitored, the set level may have been exceeded (although some "overshoot" may not be critical in an industrial situation). It is important therefore to optimise the set up of the system before using it to wash produce.

- It is essential to calibrate meters according to the manufacturers' instructions and also to maintain the probes properly. For example, the probes have electrolyte which needs to be replaced at intervals. Waste flow from the meters is returned to tank and these waste streams are appropriate for calibration of chlorine readings in that they represent what has just passed through and been measured by the meter, rather than what is present in the tank. Chlorine monitors should be calibrated at a stable reading appropriate to the levels being used. A suitable time is when the system only has potable water in it. Calibration should be against DPD readings because standard chlorine solutions cannot be purchased (see Appendix 1). pH and redox probes on the other hand can be calibrated using standard solutions.

- Aeration generates bubbles which interfere with the performance of meters unless they can be removed.

3. CURRENT PRACTICE

The following information is based on the findings of the CCFRA survey which has been published in full (Seymour 1999).

There are no standard washing techniques in the fresh produce industry. Broadly speaking, equipment is of a batch or continuous type. In batch washing, chlorinated water is used to fill tanks and vegetables are added and left for a period of time and then removed. Periodically the water is drained and replaced. In continuous systems, produce typically encounters cleaner water as it passes through the washing process, i.e. cleaner water meets the produce after it has had the gross contamination removed: this is described as countercurrent flow. Agitation with air or water may or may not be used. Water is often chilled before being used to wash produce. Although specific advice cannot be given on chilling of water because no work has been carried out at CCFRA, it is worth considering the merits and demerits of chilling in terms of the final product quality. It should be noted, however, that chlorine disinfection becomes slightly less effective as water is chilled.

Washing systems may be bought in from manufacturers or built to order by local engineering firms. Even for identical types of equipment, basic factors such as disinfectant concentration, pH control and contact time showed enormous variation. Monitoring for disinfectant and pH was also varied in nature, some using manual monitoring, some on-line and others a combination. Only 12% of respondents monitored free residual chlorine which is the most active form of chlorine as a disinfectant: this means that it is possible that washing operations are being run with no free chlorine in. Additionally, pH, another parameter significant to water disinfection, was found to be monitored by 41% and controlled by only 20% of respondents. The frequency of monitoring was also found to be highly variable.

Operators tend to judge the microbiological performance of their washing processes on the basis of Total Viable Count (TVC) reduction on their produce rather than on the basis of pathogen reduction. Pathogen reduction is hard to assess in a factory situation because the numbers of pathogenic organisms liable to be found naturally on fresh produce is small and therefore it is hard to study removal of such naturally occurring organisms.

Results from producers indicate that TVC reductions on fruit and vegetables are 100 fold at best. Calculations on TVC reduction are easier to perform using naturally occurring populations, but the statistical element still has to be taken into account and results have to be interpreted with caution due to the uneven level of contamination across the surface of the produce. In some cases, for example, there can be an apparent increase in contamination after washing, due to the vagaries of microbiological sampling.

The CCFRA survey revealed a wide range of acceptable TVC values from 10^3-10^6 per gram of product. Many buyers impose limits on microbiological quality and additionally there are guidelines provided by, for example, the British Sandwich Association[1]. Clearly it is useful to validate a washing system and take samples to monitor microbiological trends, but it is difficult to achieve a precise standard with a produce washing process as much will depend upon the initial raw material quality.

[1] BSA, 8 Home Farm, Ardington, Oxfordshire, OX12 8PN

4. PRINCIPLES OF CHLORINE DISINFECTION

Before considering the principles of chlorine disinfection, some generally used terms need explaining, as they are used widely in discussions of chlorine-based disinfection. These terms are: free chlorine, combined chlorine and total chlorine. How the different forms arise and their significance for disinfection are reviewed later in this section and in other parts of this guide. The basic definitions are presented in the following box.

Some definitions of states of chlorine

Free chlorine	$HOCl$	hypochlorous acid
	OCl^-	hypochlorite ion
	Cl_2	chlorine (gas)
Combined chlorine	$NHCl_2$	
	NH_2Cl	
	and others	

Free chlorine + Combined chlorine = Total chlorine

Sodium hypochlorite (NaOCl) when added to water produces hypochlorous acid (HOCl) and sodium hydroxide (NaOH) (Reaction 1).

Chlorine gas when added to water produces hypochlorous acid (Reaction 2). Hypochlorous acid dissociates to produce the hypochlorite ion (OCl⁻) (Reaction 3). Free chlorine reacts with other compounds to form combined chlorine compounds which are less active as disinfectants.

Reaction 1)	$NaOCl + H_2O$	$\rightarrow$	$NaOH + HOCl$
Reaction 2)	$Cl_2 + H_2O$	$\rightarrow$	$HOCl + HCl$
Reaction 3)	$HOCl$	$\rightarrow$	$H^+ + OCl^-$

The extent of the dissociation in Reaction 3 is pH dependent and the proportions of the different species present in solution are therefore affected by pH.

pH	% HOCl	%OCl⁻
5	100	0
6	97	3
7	78	22
8	23	77
9	4	96
10	0	100

Below pH 4, chlorine exists in solution as chlorine gas (Cl_2). If the pH is allowed to fall to this level, e.g. by dosing excessive amounts of acid, chlorine gas may be given off, which can be harmful to factory workers.

The hypochlorite ion is a less effective sanitiser than hypochlorous acid; however, it is still effective. At pH 8 approximately 23% of the chlorine is still in the form of hypochlorous acid.

The combination of hypochlorous acid, hypochlorite ion and chlorine gas together constitute free chlorine or free available chlorine.

Whilst in the water industry, chlorine gas is often used as a disinfectant, in the fruit and vegetable processing industry sodium hypochlorite is commonly used. Addition of sodium hypochlorite to potable water without any pH correction gives an alkaline solution due to the formation of sodium hydroxide. However, in practice the addition of fruit or vegetables to the water causes the pH to drop.

Some operations use acid (commonly citric acid) to reduce pH, whilst others do not. In these cases (where acid is not used) the addition of vegetables will reduce pH, but the pH may not always fall to the optimal range for disinfection. A target pH would typically be 7.0-7.5. Low pH values give rise to more corrosion. Use of an alkaline solution to adjust pH should not normally be necessary.

Free chlorine should be distinguished from combined chlorine, which is formed when free chlorine reacts with ammonia or organic nitrogen-containing compounds. Combined chlorine compounds (e.g. chloramines) are much less effective as disinfectants than free chlorine although they do exert an effect. Free chlorine and combined chlorine together constitute total chlorine.

Under certain conditions, total chlorine can be depleted as chloride ions are formed from chloramines.

Key points:-

- Free chlorine is the most effective form of chlorine for disinfection.

- Free chlorine can be generated from chlorine gas, but in the fresh produce industry sodium hypochlorite tends to be used.

- Free chlorine is more effective as a disinfectant below pH 8 as hypochlorous acid (HOCl) becomes more prevalent.

- Free chlorine reacts with other compounds to form combined chlorine compounds which are not such effective disinfectants.

5. ORGANIC LOADING - PRINCIPLES AND CASE STUDY

Addition of organic material to chlorinated water causes an immediate drop in the free chlorine level due to the reaction of free chlorine with organic nitrogen-containing compounds or ammonia (the so-called "chlorine demand").

When hypochlorite is added to a tank of potable (e.g. mains) water, all the chlorine generated will be free chlorine due to lack of chlorine demand of this water. If produce is then added and consequently the water contains organic compounds, additional hypochlorite does not just generate free chlorine but also combined chlorine. Sufficient hypochlorite therefore has to be added to ensure that free residual chlorine is present as well. As the water becomes more and more contaminated with further batches of produce, the levels of some of these combined chlorine compounds may be enough to cause tainting. Two options exist to prevent this: in a batch system the water may be replaced sufficiently often to prevent this occurring; whilst in a continuous system an appropriate solution is to top up the tank with mains make-up water. In this case, combined chlorine products are being diluted on a continuous basis and the excess volume can be re-circulated to be used, for example, in a pre-wash tank.

Case Study - effect of organic loading

This case study looks at various aspects of the organic load added to washing systems and the effect that it has on the free chlorine level and pH. It was intended to demonstrate

1. The effect of product:water ratios on the reduction of free chlorine in the tank
2. Any variation between product types in terms of chlorine demand
3. Any differences between coarse cut and fine cut produce
4. Any differences between product type in terms of measured removal of microorganisms

In all cases experiments were conducted in a 300 litre wash tank, where different amounts of vegetables, from 1-25 kg, were added to a pH controlled system at pH 7.0 and readings were taken of changes in free and total chlorine and also pH.

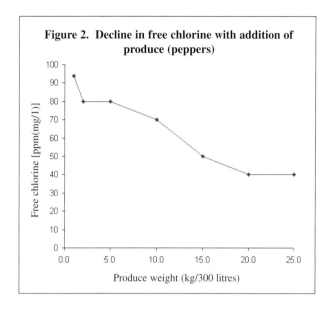

Figure 2. Decline in free chlorine with addition of produce (peppers)

Product:water ratio

Figure 2 illustrates the effect of increasing the ratio of product to water on the level of free chlorine left in the tank for one produce type (peppers).

Addition of between 1 and 25kg of product to the 300 litre tank showed the increasing effect of organic load on free chlorine, i.e. the more produce added, the greater the drop. Addition of the greatest amount (25kg) did not remove free chlorine completely.

Figure 3. Decline in free chlorine with different produce types

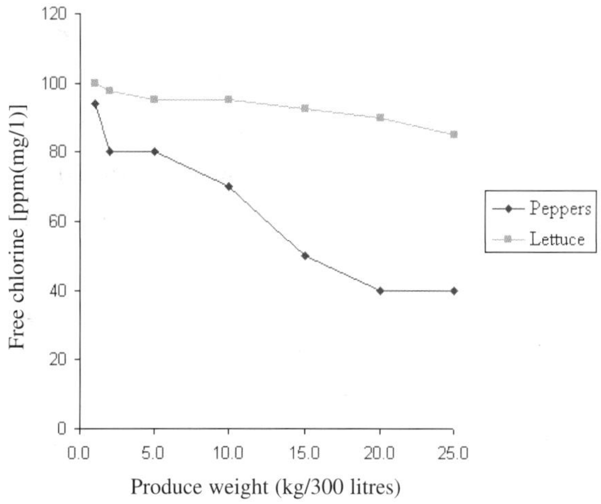

Chlorine demand of different product types

A variety of vegetables was tested, namely pepper, lettuce, cabbage and carrots. Differences were found between the vegetables and Figure 3 illustrates the difference between peppers and lettuces which have different surface areas and structures.

Coarse cut versus fine cut produce

It would be anticipated that the leakage of cell contents from produce would lead to greater chlorine demand with finer slicing of vegetables. From the work carried out at CCFRA the evidence for this was not clear, but it is worth considering. The difference between whole and sliced produce is likely to be the most marked.

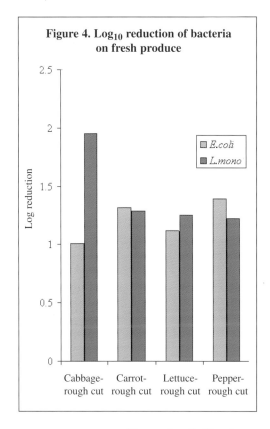

Figure 4. Log$_{10}$ reduction of bacteria on fresh produce

Decontamination of vegetables spiked with microorganisms

Decontamination work was carried out on vegetables (pepper, lettuce, cabbage and carrots) spiked with microorganisms and introduced to the washwater (with air agitation) after the drop in free chlorine had been measured. The initial free chlorine level was 100ppm in all cases, but this fell to 40-80ppm after addition of 25kg of produce to the tank. *Listeria monocytogenes* and *Escherichia coli* were spiked onto vegetables and recovered using a method described in Appendix 2. The reduction in numbers (log$_{10}$) was calculated (Figure 4).

In all cases, decontamination was achieved in the range 90-97% (1-1.5 log$_{10}$) although rough cut cabbage showed nearly 2 logs removal of *Listeria monocytogenes*. Reproducibility between replicates was good. This illustrates that decontamination was occurring despite the organic load. Reuse of such wash water with a second batch of vegetables could, in the case of peppers, completely eliminate free chlorine. In the case of lettuce, the demand is not so great and the water can clearly be used more than once. This sort of information can be used to help plan washing regimes.

Often a figure is quoted for chlorine level in a batch washing system. In general, this is the initial free chlorine level rather than the level left at the end. In a batch system it is clearly important to retain a measurable amount of free chlorine at the end of the washing process to guarantee that chlorine has been present throughout the process.

Key points

- The amount of chlorine left at the end of the wash is a function of its initial concentration and the amount and type of organic material being washed.

- Operators should establish the chlorine demand in their particular system and therefore the number of washes of particular produce types per batch of water.

- Checks are required to ensure that the starting concentration of chlorine is correct.

- A fail-safe system is required to demonstrate that free chlorine is present at the end of a wash.

- Decontamination at pilot scale was shown to be fairly consistent across the types of produce tested.

6. CONTROL OF CHLORINE AT LOW LEVELS: PRINCIPLES AND CASE STUDY

Rather than adding large amounts of free chlorine at the start and disposing to waste the water when that chlorine is lost, it is also possible to retain a residual level of chlorine with continuous dosing. In the case described below, the system was deliberately overloaded and the chlorine then allowed to recover. In factory situations, with the correct control of produce loading rate and chlorine dosing rate, it will be possible to retain a free chlorine residual. This represents an alternative way of managing a disinfection system.

Case Study: control of free chlorine at 5 and 10ppm

This case study was intended to look at relatively low free chlorine residuals compared to those currently quoted in the industry and to compare the microbial reduction under these conditions with that when free chlorine has been lost completely. The principle was to obtain a level of 5 or10ppm free chlorine at pH 7.2, add 15 kg of cabbage to remove free chlorine, and allow the level to decline and then recover. It was possible to compare decontamination when the chlorine had been lost to when it had recovered. Although the control system tried to maintain lower levels of chlorine (5 and 10ppm), it would equally apply to higher levels.

It was intended to demonstrate:

1. How addition of vegetables caused a free chlorine level (set at 5 or 10ppm) to drop to zero
2. How the control system allowed recovery to the set level
3. The relative decontamination of cabbage at these two levels of chlorine
4. Any effects of time on sanitisation

With the 300 litre tank just containing potable water (and with no agitation), the free chlorine monitor was set to either 5 ppm or 10ppm and the pH to 7.2. Free chlorine, total chlorine and pH were the parameters measured. Initially the free chlorine increased up to the set level (Figs 5 and 6). A decline in free chlorine on addition of organic matter led to its loss within 6 minutes. Recovery then ensued as the monitoring/dosing system compensated for this loss. It would have been possible to have higher rates of chlorine dosing, but this would have led to overshoot of the set targets. It is not always easy with a control system to ensure that set points are met without being exceeded.

Changes in total chlorine levels followed those observed with free chlorine, suggesting that combined chlorine compounds formed were breaking down to form chlorides (see Section 4). No major pH changes were noted, presumably because the pH of the cabbage was not low enough to have an effect.

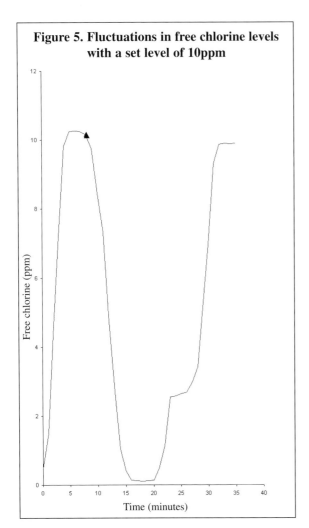

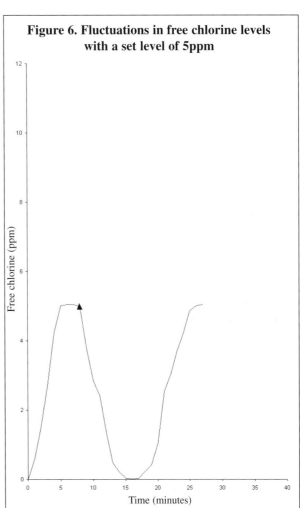

The triangle symbol indicates the point at which the 15kg of cabbage was added

Additionally, decontamination experiments were carried out with specific inoculated microorganisms and using the method described in Appendix 2 (Figures 7 and 8). For both 10ppm and 5ppm set values, greater decontamination was achieved when the chlorine level was at these levels than when it had dropped to zero. Although this is to be expected, it is useful to illustrate the point that a small free chlorine residual is more effective than a zero level.

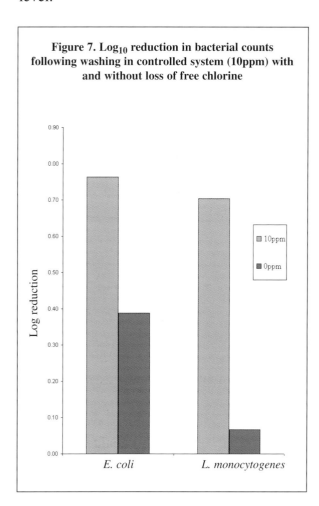

Figure 7. Log_{10} reduction in bacterial counts following washing in controlled system (10ppm) with and without loss of free chlorine

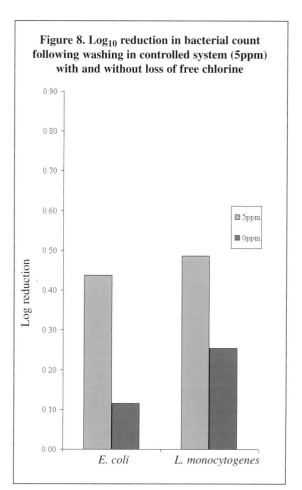

Figure 8. Log_{10} reduction in bacterial count following washing in controlled system (5ppm) with and without loss of free chlorine

Decontamination work was also carried out for different time periods and the effectiveness of increasing wash times was illustrated (Figures 9 and 10). Some increase in removal is observed with time although the amount of additional decontamination gained with a 10 minute rather than 5 minute wash was minimal.

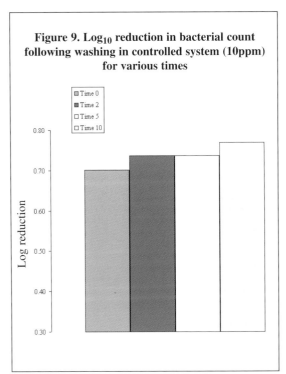

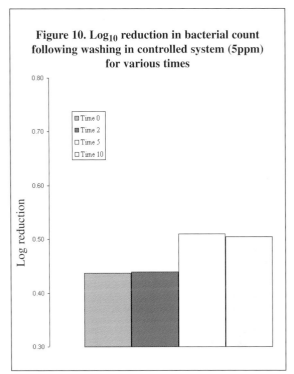

* note: time 0 had minimum contact time, i.e. ≤5s

Key Points:

- Free chlorine levels may be controlled at low levels, but may be lost from the system due to addition of organic load.

- The system should be designed to recover quickly or a higher set point should be chosen to avoid zero chlorine levels.

- Microbial decontamination is noticeably better with a low level of chlorine than with no chlorine with the procedure described.

7. WHAT TO MONITOR: FREE CHLORINE, TOTAL CHLORINE OR REDOX POTENTIAL?

Free chlorine and pH

Free chlorine should be measured in vegetable washing systems, at least by manual means. The combination of free chlorine measurement with a pH reading gives an accurate assessment of the disinfective capability of free chlorine. Automated measurement of these parameters is also a possibility. This may or may not be linked to dosing equipment through feedback control.

If automated measurement of free chlorine is undertaken, it is important to calibrate the on-line monitor on a regular, preferably daily, basis against colorimetric DPD readings (see Appendix 1).

Total chlorine

Total chlorine measurement takes into account both free chlorine and combined chlorine. As combined chlorine is a much less effective disinfectant than free chlorine, measurement of total chlorine is useful only under certain circumstances. For example, if there is no combined chlorine in the system, then total chlorine measurements will match free chlorine measurements: this will be the case if a washing tank is being filled with potable grade water. As combined chlorine is generated, total chlorine levels will give an indication of the levels of these potentially taint-causing chemicals.

Measurement of total chlorine may be undertaken either with a colorimetric DPD based system or with a calibrated total chlorine probe.

Redox

A redox measurement takes into account the ratio of the oxidants to reductants present in the washing tank. In a tank with potable water and hypochlorite, the ratio between free chlorine and chlorides will cause redox to follow free chlorine levels.
Addition of organic material complicates this picture and it is not possible to reliably use redox as a measure of free chlorine under these circumstances.

As combined chlorine compounds build up, the increase in concentrations of reduced chlorine compounds allows the redox to act as a measure of combined chlorine present.

Key points

- Free chlorine and pH should always be measured

- Total chlorine or redox may be used to indicate levels of combined chlorine compounds as wash water gets more and more contaminated with organic material

8. ALTERNATIVES TO CHLORINE

The fresh produce industry in the UK would ideally like to see an alternative to chlorine, which was easy to use and seen as a "friendly" product in terms of food use but which is also highly effective as a disinfectant or sanitiser. Some companies have moved away from chlorine already.

A major issue is an understanding of the natural contamination found on fresh produce and the ways that it can most easily be removed and this will require fundamental research. There are some established alternatives to chlorine which have been, in general, less well characterised than chlorine itself but have some features to recommend them.

Other oxidative bioxides

Chlorine dioxide

Chlorine dioxide (ClO_2) is recognised as an effective water disinfectant. It is equally effective across a wide pH range, unlike chlorine, but because it is an oxidant it will still react with organic material in the water and will therefore need to be replenished periodically.

Chlorine dioxide is generated on site from precursor chemicals. Increasingly this is a simple process with the on-site health and safety risks being minimised. It is used in a range of food industry applications including potato washing and is becoming more widespread for fresh produce decontamination. To a large extent, its use is probably behind chlorine for historical reasons. Beauchat (1998) notes the variability of ClO_2 in killing microorganisms on surfaces. It is not known, however, how this relates to fresh produce, and more detailed studies are needed comparing ClO_2 with chlorine.

More information on ClO_2 may be found in CCFRA Guideline No.15 (Holah, 1997)

Ozone

Ozone is increasingly popular for drinking water disinfection. It is also used in some swimming pools and gives the advantage of not being associated with similar by-products to chlorine. It exists as an unstable gas and rapidly decomposes to O_2 with the rate of decay being dependent upon the purity of the water. Ozone must be generated on site and there are some health and safety considerations associated with its use that can be overcome. As with chlorine dioxide, there is a lack of published information comparing ozone and other methods of disinfection in fruit and vegetable decontamination.

Hydrogen peroxide

Hydrogen peroxide (H_2O_2) is an oxidising compound supplied as a liquid which is sold for various disinfective purposes; it breaks down to water and oxygen gas and is relatively safe in its application. Its use as a decontaminant of fresh produce has not been widely recognised, although some studies have been carried out and more work may be justified (Beauchat, 1998). This disinfectant has been associated with browning of shredded lettuce and also bleaching of the surfaces of produce.

Peracetic acid

Peracetic acid ($CH_3 COOH$) is a powerful oxidant as well as being an organic acid. Peracetic acid solutions sold commercially are mixtures of peracetic acid, hydrogen peroxide, acetic acid and water. Such solutions are stable, but they decompose to acetic acid and hydrogen peroxide when diluted. It has been used to treat seeds, but little published work has been carried out with fresh produce.

Organic acids

Organic acids (i.e. those which are compounds of carbon) may be used to retard growth or kill microorganisms. Unfortunately in many cases, treatment with these acids may change the flavour or aroma of products and so only be suited for products subsequently incorporated into flavoured salads. When these acids have been used, considerable reductions in log count have been demonstrated (see for example Karapinar and Gonul, 1992). Such treatments have been used in combination with chlorine, e.g. acid dip first followed by chlorine dip. The use of organic acid or vinegar treatment is not uncommon in the industry. The acids used include lactic acid, citric acid and acetic acid. It should be noted that these compounds contribute significantly to the biochemical oxygen demand (BOD) of waste water.

Surfactants

A surfactant is a molecule which has both water soluble and water insoluble parts to it. These properties allow it to lower the surface tension of water, thereby improving surface wetting. Potentially surfactants could improve physical removal of microorganisms or allow better penetration of disinfectants. They are included in some proprietary washes, but their potential use has not been fully evaluated.

9. CONCLUSIONS

Chlorine is used by a large proportion of the UK industry in fresh produce washing and its use is also common worldwide. There is no immediate indication that it would be banned for fresh produce washing, particularly as washing is often the only step in the removal of potentially harmful bacteria. It clearly gives more removal than potable water alone and is also well accepted by the water industry as a disinfectant, although it is not seen as particularly desirable by all sectors of the food industry.

Chlorine disinfection requires understanding of the principles described in this guideline to maximise its efficient use.

Key issues for consideration

- A measurable level of free chlorine is required at the end of a washing process

- Free chlorine does not have to be at very high levels to give better disinfection than potable water

- Careful consideration is required of the effect of the produce on chlorine levels and therefore the effect on the washing regime

10. ACKNOWLEDGEMENTS

We wish to acknowledge the following organisations and people:

Prominent Fluid Controls for provision of equipment and comments on the text by Jane Cooper, Andrew Armytage and Bob McAlster.

Graham Christie of RHM Technology and Debra Davies of Westward Laboratories, who read and commented on the draft.

Colleagues at CCFRA.

11. REFERENCES

Anon (1996). Guidelines for drinking water quality. Volume 2: Health criteria and other supporting information. World Health Organisation. International Programme on Chemical Safety.

Anon (2001) *Salmonella* Newport infections associated with the consumption of ready-to-eat salad. CDR Weekly, **11** (6), www.phls.org.uk

Beauchat, L.R. (1998) Surface decontamination of fruits and vegetables eaten raw: A review. Food Safety Unit, World Health Organisation *WHO/FSF/FOS/98.2*

Holah, J. (ed.) (1997) Microbiological control of food industry process waters: Guidelines on the use of chlorine dioxide and bromine as alternatives to chlorine. CCFRA Guideline No. 15

Karapinar, M. and Gonul, S.A. (1992). Removal of *Yersinia enterocolitica* from fresh parsley by washing with acetic acid or vinegar. International Journal of Food Microbiology, **16**, 261-264.

Lash, S. (2002) FDA finishes produce survey, finds pathogens in 12 samples. Food Chemical News, **43** (50), 18

Little, C., Roberts, D., Youngs, E., and de Louvois, J. (1999) Microbiological quality of retail imported unprepared whole lettuces: A PHLS food working group study. Journal of Food Protection, **62** (4), 325-328

Seymour, S. (1999) Review of current industry practice on fruit and vegetable decontamination. CCFRA Review No. 14

Sommer, P., Martin-Rouas, C. and Mettler, E. (1999) Influence of the adherent population level on biofilm population, structure and resistance to chlorination. Food Microbiology, **16**, 503-515

Zhang, S. and Farber, J.M. (1996). The effects of various disinfectants against *Listeria monocytogenes* on fresh cut vegetables. Food Microbiology, **13,** 311-321

Appendix 1

Procedure for the Measurement of Free Chlorine using the DPD method

The DPD method is the standard way of evaluating free chlorine concentration in water.

N,N-diethyl-p-phenylenediamine (DPD) is a reagent which develops a stable pink coloration in contact with oxidising agents.

Kits are available which operate on the basis of comparison against a standard colour wheel known as a comparator. This comprises a gradation of coloured "windows" representing different chlorine concentrations. It is possible to purchase comparators with different ranges of sensitivity.

It is also possible to measure the colour generated by the DPD method in a dedicated hand held spectrophotometer. These are more costly than comparators but some operators prefer to use them because they take a reading automatically. Used properly, either method is satisfactory for the purposes of the food industry. Plastic cuvettes are available to avoid the risk of glass being present in the production area.

Test strips are also sold for chlorine measurement. These are simply dipped in water and the colour read off a chart. They are not as accurate as comparators or spectrophotometers but give a very crude measure of chlorine which may be useful under some circumstances.

A typical DPD comparator method for free chlorine is described below: however in all cases the specific manufacturer's instructions should be followed. This method is for a low range kit: high range kits are also sold.

The test kit will only measure up to 6ppm of chlorine. Therefore if a higher level of chlorine is to be measured, a dilution must be carried out, in order to obtain an accurate reading. For example if you are testing a sample that is likely to be about 20ppm of chlorine, a 1:10 dilution will be required, but for a sample that is likely to be about 200ppm of chlorine, a 1:100 dilution is required. For the 1:10 dilution, mix 1ml of sample with 9ml of deionised water; for a 1:100 dilution, mix 1ml of sample with 99ml of deionised water. Note that if a dilution has been carried out, the free chlorine reading obtained must be multiplied by the dilution factor.

1. Rinse a 13.5mm square cell with the sample to be tested, leaving 2 or 3 drops in the cell.
2. Add one DPD No.1 tablet and crush it in the cell with the plastic stirring rod
3. Fill the cell to the 10ml mark with the sample and mix well to dissolve the tablet. (A pink colour should result)
4. Fill another cell with 10ml of the sample.
5. Fit the free chlorine disc into the comparator, and place the cell, containing sample only, into the left compartment and the cell containing the sample and DPD tablet into the right compartment.
6. Hold the comparator facing natural light, preferably northern daylight, and rotate the disc until the nearest colour match is obtained. The reading is taken from the window in the bottom right hand corner of the comparator: this is the free chlorine reading. Remember to multiply this reading by the dilution factor used.
7. Rinse the 2 cells and stirring rod thoroughly with clean water.

Note: if, when the DPD tablet is first crushed, a red colour is produced, which then bleaches to colourless when the remaining sample is added, there is a high concentration of chlorine in the sample. The sample should be diluted and the test repeated.

Appendix 2

Method for spiking and recovering microorganisms from vegetables

Single (ampicillin resistant) colonies from nutrient agar with 0.1% ampicillin were inoculated into nutrient broth (with 0.1% ampicillin) and incubated for 24h at 37°C (*E. coli*). Single colonies from nutrient agar were inoculated into TSBYE and incubated for 24h at 30°C (*L. monocytogenes*).

Concentrations of bacteria were determined using optical density measurements based upon standard curves obtained previously and the inoculum used was approximately 10^8/ml in the relevant broth.

100g of product was placed into a plastic Model 400 Seward Stomacher Bag with 1ml of inoculum, and then heat sealed. Each bag was shaken gently 30 times to distribute the inoculum and then the samples were stored overnight at 4°C. 2 x 10g of product was analysed to determine inoculum levels. This method follows that of Zhang and Farber (1996).

Enumeration of bacteria

Samples (10g) of fresh produce were diluted with 90ml of maximum recovery diluent (MRD) and then homogenised in a stomacher (Lab-blender 400, Seward, UK) for 120s. Serial dilutions were then prepared in MRD and microorganisms enumerated by plating onto selective media. In the case of *L. monocytogenes*, spread plates were prepared using *Listeria* Selective Agar, Oxford formulation (Oxoid CM856) plus *Listeria* Selective Supplement (Oxoid SR140E). Incubation was at 30°C for 48h. In the case of *E. coli*, pour plates were made using Nutrient Agar (Oxoid, CM3) with 0.1% ampicillin. Incubation was at 37°C for 24h.